EXPLORING
NEPTUNE

By Emma Jones

KidHaven PUBLISHING

Published in 2018 by
KidHaven Publishing, an Imprint of Greenhaven Publishing, LLC
353 3rd Avenue
Suite 255
New York, NY 10010

Designer: Deanna Paternostro
Editor: Vanessa Oswald

Photo credits: Cover, back cover, p. 19 (main) Vadim Sadovski/Shutterstock.com; pp. 4–5 Naeblys/ Shutterstock.com; p. 7 D1min/Shutterstock.com; p. 9 Orange-kun/Wikimedia Commons; p. 11 NASA images/Shutterstock.com; p. 13 Dorling Kindersley/Getty Images; p. 15 Supportstorm/Wikimedia Commons; p. 17 PlanetUser/Wikimedia Commons; p. 19 (inset) Time Life Pictures/Contributor/The LIFE Picture Collection/Getty Images; p. 21 Elena Duvernay/Stocktrek Images/Getty Images.

Library of Congress Cataloging-in-Publication Data

Names: Jones, Emma, 1983- author.
Title: Exploring Neptune / Emma Jones.
Description: New York, NY : KidHaven Publishing, [2018] | Series: Journey through our solar system | Includes bibliographical references and index.
Identifiers: LCCN 2017003830 (print) | LCCN 2017009930 (ebook) | ISBN 9781534522831 (pbk. book) | ISBN 9781534522572 (6 pack) | ISBN 9781534522725 (library bound book) | ISBN 9781534522602 (eBook)
Subjects: LCSH: Neptune (Planet)–Exploration–Juvenile literature.
Classification: LCC QB691 .J66 2018 (print) | LCC QB691 (ebook) | DDC 523.48–dc23
LC record available at https://lccn.loc.gov/2017003830

Printed in the United States of America

CPSIA compliance information: Batch #BS17KL: For further information contact Greenhaven Publishing LLC, New York, New York at 1-844-317-7404.

Please visit our website, www.greenhavenpublishing.com. For a free color catalog of all our high-quality books, call toll free 1-844-317-7404 or fax 1-844-317-7405.

CONTENTS

FAR-OUT PLANET

Neptune is the farthest planet from the sun in the **solar system**.

Mercury

Venus

Earth

Mars

It's the only planet that can't be seen without a **telescope**. Neptune is 2.8 billion miles (4.5 billion km) from the sun.

Jupiter

Saturn

Uranus

Neptune

Neptune is the fourth-largest planet.

It takes Neptune about 165 Earth years to **orbit** the sun one time! After its discovery in 1846, Neptune didn't complete its first full orbit until July 12, 2011.

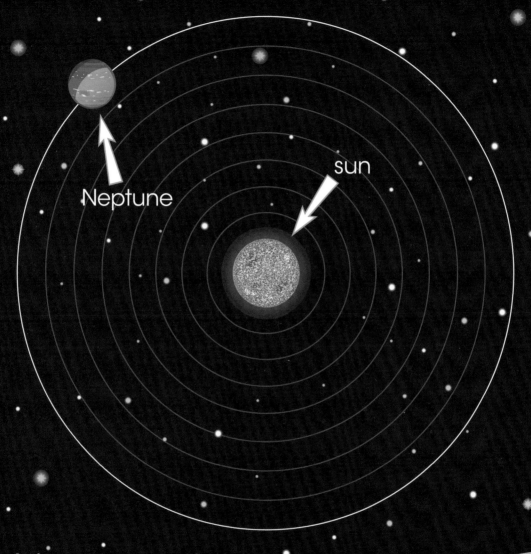

Neptune

sun

Astronomers used math
to discover Neptune.

PLANET OF GASES

Neptune is known as a gas giant. The planet's **atmosphere** is made up of thin clouds. Beneath the clouds is a thick **layer** of gases. Neptune's ground is not solid like Earth's.

People couldn't live on Neptune
because it doesn't have solid ground.

Neptune's thin, cloudy layer has different parts. These parts spin around at different speeds. The top and bottom parts make a full spin in 12 hours. The middle part takes 18 hours to make a full spin.

The spinning of Neptune is also known as its rotation.

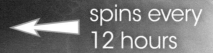
spins every 12 hours

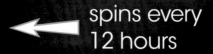
spins every 18 hours

spins every 12 hours

NEPTUNE'S LAYERS

The top layer of Neptune is made up of gases such as methane, hydrogen, and helium. Beneath this is an icy layer of water, **ammonia**, and methane. Methane is what makes Neptune look blue. The center layer is made of ice and rock.

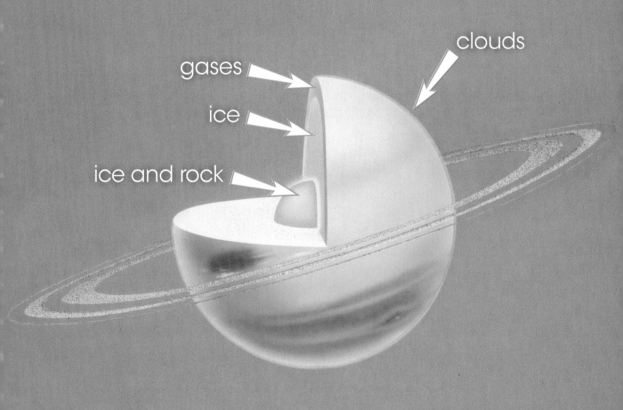

clouds

gases

ice

ice and rock

The center, or core, of Neptune is about the same size as Earth.

STORMY WEATHER

Neptune is the planet of bad weather! Dark spots on the planet are big storms. Neptune has the strongest winds in the solar system. They are about nine times faster than winds on Earth!

Neptune's average temperature
is -353 degrees Fahrenheit
(-214 degrees Celsius).

storms

The Great Dark Spot was a large storm on Neptune. The storm was about the size of Earth. Winds at speeds up to 1,500 miles (2,414 km) per hour were recorded. These are the strongest winds ever recorded on any planet in the solar system!

Great Dark Spot

The *Voyager 2* **probe** found the Great Dark Spot in 1989. This storm could no longer be seen a few years later.

MOONS AND RINGS OF NEPTUNE

Neptune has 14 moons. The largest moon is named Triton. The *Voyager 2* probe found six moons in 1989. The last moon was discovered in 2013 in old telescope pictures, which had been lost for more than 25 years.

Triton has geysers that spray icy chunks more than 5 miles (8 km) into its cold atmosphere.

The *Voyager 2* probe discovered Neptune has six thin rings that are hard to see. These rings are all different sizes. There are thick areas in the rings called arcs. These arcs are made of dust.

Other planets with rings are Saturn, Jupiter, and Uranus.

rings

GLOSSARY

ammonia: A colorless gas with a strong smell.

astronomer: A person who studies different parts of the solar system.

atmosphere: Gases in the air around a planet.

layer: One part of something lying over or under another.

orbit: To travel in a circle or oval around something.

probe: A vehicle that sends information about an object in space back to Earth.

solar system: The sun and all the space objects that orbit it, including planets and their moons.

telescope: An instrument used to view distant objects, such as the planets in space.

FOR MORE INFORMATION

Websites

ESA Kids: Neptune

www.esa.int/esaKIDSen/SEM7CTMZCIE_OurUniverse_0.html

This website includes fun facts about Neptune.

NASA Space Place: All About Neptune

spaceplace.nasa.gov/all-about-neptune/en/

Visitors to this website can see pictures of Neptune as they learn about the planet.

Books

Adamson, Thomas K. *The Secrets of Neptune*. North Mankato, MN: Capstone Press, 2016.

Roumanis, Alexis. *Neptune*. New York, NY: AV2 by Weigl, 2016.

INDEX